Drone

FAA Part 107 License Practice Test Questions & Answers For Seniors

Over 180 Test Questions and Answers to Ace Your Part 107 License Test at First Attempt

ROBERT
GONZALO

Copyright

Printed in the United States of America

Table of Contents

CHAPTER ONE

Introduction to the Drone FAA Part 107 License

Drones are currently flying high in several parts of the states. Drone Evaluations in 2018 reveal that the global drone market grew by up to 42%, and a progression of 14,300 million is estimated in 2025. The advancement in drone technology has led to its introduction into agriculture, mechanical inspections, emergencies and security on a large scale.

Drone technology has now become an accessible and lucrative business globally. However, those who want to dedicate themselves to professional drone flight need to meet a series of requirements, including drone flight licenses. The Federal Aviation Administration (FAA), a body in charge of regulating aviation in the US, lays specific regulations concerning the drone's law.

These rules and regulations are entirely based on the use of the drone, e.g., for recreational or business purposes. If your goals are recreational or entertainment, it is unnecessary to obtain a drone pilot certificate. However, this does not mean that getting a license is wasteful.

The devices that fall within the recreational purpose should also not exceed 0.55 Lbs. However, if a drone is to be used within a working environment, it is necessary to obtain a drone license. Therefore, studying and preparing for the FAA test becomes paramount. Additionally, studying for the test will widen your scope on the various aspects related to the use of these aircraft, regulations, and maintenance practices.

Do you need a license to fly your drone?

The answer is NO if you are flying a recreational drone. However, the pilot needs to comply with the following rules:

- The pilot must fly his drone at a minimum distance of 8 km from any airport or airfield.
- You cannot fly within controlled airspace.
- The pilot cannot pass his drone more than 120 meters high from the ground or higher obstacles

within a 150-meter radius from the drone.

- You cannot fly an unmanned drone weighing above 0.55Lbs. In addition, you will not be allowed to fly in urban areas and over people, and above 20 meters in height.

What is an FAA Part 107 Test?

Part 107 is a peculiar test directed by the Federal Aviation Agency to train and control small unmanned aircraft. After passing the part 107 test, the pilot will be granted a license to operate a drone. The test is a means of ensuring safety for both the pilot and the people out there.

What does the Part 107 Test look like?

A glance at the complete list of topics that will be asked in the Part 107 test:

1. **Regulations:** This segment concerns the FAA's laid down instructtions and guidelines for safely operating small-unmanned aircraft.
2. **Airspace classification:** This segment is concerned about the different classifications that exist in the airspace. This classification will

enlighten the pilot to know which airspace to fly.

3. **Weather:** This segment is concerned with the adverse effects of the elements on the drone and the measuring instrument for conducting weather checks.
4. **Loading:** This segment is concerned with the load required to balance the drone properly before flight. It deals with the centre of gravity as well as the thrusting weight of the drone.
5. **Emergency:** This segment is concerned with the pilot's action when an inflight emergency occurs.

6. **FUI:** This segment is sometimes classified with the regulation section. However, FUI means Flying under the Influence; it concerns the materials consumed by the pilot-in-charge before controlling the small-unmanned aircraft.
7. **Decision Making:** This segment is concerned about frequent decision pilot make while operating small-unmanned aircraft.
8. **Maintenance:** This segment is concerned with the care of the drone as well as the scheduled protocol of maintenance from the

manufacturer and the pilot-in-charge.

It is important to note that the sixty (60) questions of the Part 107 exam are objective with multiple options. However, the FAA requires you to have a complete understanding of every topic listed above to score above 70% on your test.

How much does obtaining a drone license cost?

The cost of getting your drone license is not as high as people think. The price is less than $200. However, it is essential to note that this price ($200) does not cover the fee for a study guide or test guide to ace your study.

Where is the FAA drone Test Center?

If you have registered to participate in the testing, there are about 800 locations in the US states and territories. The choice of a centre is dependent on the closest area to your apartment.

In conclusion, drone technology is here to stay, as the constant evolution of its models has opened a world of great financial possibilities for those who want to dedicate themselves to the use of professional drones. First, you need to pass the part 107 exam with excellence to get a license.

CHAPTER TWO

New Changes to the Part 107 Rules

December 2020, FAA released certain modifications to the part 107 rules. These changes are aimed to ensure proper guidance of the aircraft and impact many transformations to the Part 107 certificate holders in the future. FAA new changes a documented in a 750 pages document, which can be summarized as follow;

1. **Operating Rules for the Unmanned Aircraft (UA):**

- All UA registered for commercial purposes are required to have a standard remote ID. However, this rule does not apply to UA used for recreational purposes except it weighs above 0.55 lbs or serves other commercial purposes.

- The broadcast remote ID serves the purpose of transferring messages to the different control centres. The Standard Remote ID message includes the ID, current location, altitude, speed, emergency status, etc., of the unmanned aircraft.
- A separate remote ID broadcasting module can be attached to the unmanned aircraft. This feature allows the use of an in-built broad-

cast module, and its serial number must be entered into the registration record for the unmanned aircraft.

- Recognized identification area from the FAA: This region is only allowed by UA that is not adequately equipped with the remote ID. Organizations and certain establishments can apply for this FRIA.

2. **Design and Production Rules for Manufacturers:**

- The recent production of the unmanned aircraft must contain the standard remote ID

- The manufacturer of the UA must produce a broadcast module that met the performance requirements of the rule.

3. **Operation of the unmanned aircraft over People**

- The FAA has banned the use of Part 107.39 waiver. The permission of drones to move over the head of people and moving vehicles is now prohibited. In addition, drone users can now operate their devices at night under certain conditions without the permission of a waiver from the FAA.

- The FAA also divided the category of drone usage into four different parts. Each class has specific rules that govern those concerning waivers they can enjoy.
- According to the FAA, the drones that fall under category one are drones that cannot cause injuries or damage to people. This type of drone weighs below 0.55Lbs. This category of drone users no longer requires FAA certification nor means of compliance (MOC) as long as the pilot is operating with a propeller guard.

- The second category of drones is those drones that can cause an injury to people or property by transferring 11 foot-pounds of kinetic energy upon impact. The FAA recommends that these drones be free of material that may cause damage to the skin of people. However, the usage of this category of drone needs the certification of FAA or means of compliance (MOC). There is a ban on the use of this category of drones over open-air assemblies. The only exception is that this drone complies with the requirement for standard remote ID.

- The third category of drone users are drones that can cause injury to people or property by transferring 25 foot-pounds of kinetic energy upon impact. The FAA recommends that these drones be free of material that may cause damage to the skin. However, the usage of this category of drone needs the certification of FAA or means of compliance (MOC). There is a ban on the use of this category of drones over open-air assemblies. The only exception is that this drone complies with the two requirements for standard remote ID.

- The fourth category of drone users requires serious measures due to their impact on lives and properties. This category of drone user requires an airworthiness certificate and regular inspection as enshrined in Part 21 of FAA regulations.

4. **Operation of the unmanned aircraft over Vehicles**

- A drone's movement above a vehicle can only be done if it falls under the following category (one, two, and three). However, the drone is not under strict adherence to sustain a flight over a moving vehicle.

- The category four drones require certification and clearance before they can move ahead of vehicles.

5. **Inclusion of the Remote Knowledge Test**

- These new changes to the former rules enable the pilot to undergo particular tests such as the night test. This test has been programmed to be recurrent every twenty-four months. It is, however, an online test, and it comes at no expense or fee to remote pilots.

6. **Inspection of Pilot and Compliance**

- The new rules that govern pilots are as follows;
  1. For commercial purposes, the pilot in charge must present their pilot certification as well as the UAS rating and identification
  2. Ensure that every document that is needed before piloting any goods under government surveillance is given to the FAA

3. There should not be an argument with officials upon the request for an inspection of a drone.

CHAPTER THREE

Test Questions on Part 107 Regulations

1. At what age is an owner of a small-unmanned aircraft not eligible for the registration?

a) When the owner of the small unmanned aircraft is not up to thirteen years of age

b) When the owner of the small unmanned aircraft is an indigene of the country

c) When the owner of the small unmanned aircraft lacks a valid means of identification

2. When is it compulsory for the owner of a small-unmanned aircraft to register with the FAA?

a) When the small unmanned aircraft weighs above .55 pounds irrespe-ctive of the use (commercial or recreation)

b) When small unmanned aircraft are used for commercial purposes

c) When the aircraft crossed a certain limitation in the airspace.

3. Where is the location according to FAA the serial number (standard remote identification or a broadcast

module) of the small-unmanned aircraft be listed.

a) It should be listed in the document of compliance of the aircraft
b) It should be listed in the manufacturer's method of compliance
c) It should be listed in the certificate of the aircraft registration

4. In what manner can a small-unmanned aircraft be operated?

a) It should be operated with care to avoid endangering of life and property of the populace
b) It should be operated with the aid of an observer

c) It should be operated to fly in a specific region of the airspace less than 200 feet AGL

5. What precaution is to be taken if you plan to release a golf ball from an unmanned aircraft with an altitude of 100 ft AGL?

a) Ensure that the release will not lead to the destruction of properties or becomes hazardous

b) Ensure to land the small unmanned aircraft in the landing zone before releasing the golf

c) Release the golf when it would not destroy property above three hundred dollars

6. What should become a pilot next course of action after having a dinner with wine with a client?

a) Administer a self-test before flying the small unmanned aircraft
b) Cease from operating the small unmanned aircraft for about 8 hours in case an alcoholic beverage is consumed
c) Pilot the small unmanned aircraft when you are not drunk

7. After the arrival of the sunset in the morning, what should a pilot controlling the small-unmanned aircraft do?

a) Put on an anti-collision light after the appearance of a sunset.
b) Move the small unmanned aircraft in a rural region
c) Ensure to use a transponder

8. What should you do as a pilot if you observe a hot air balloon when flying the small-unmanned aircraft?

a) Move to the right-of-way away from the hot air balloon

b) Control the small-unmanned aircraft to the side, below, or above the hot air balloon.

c) Wait until the hot air balloon is above the small-unmanned aircraft.

9. What is the minimum ground speed stipulated by the FAA when piloting the small unmanned aircraft?

a) 80 knots (100 mph)

b) 80 mph (75.69 knots)

c) 100 knots. (114mph)

10. When requested by the FAA, a pilot must ensure to provide ________

a) Income logbooks of the past earnings of the small unmanned aircraft

b) The remote pilot certification as well as the rating of the small unmanned aircraft

c) The pilot employers identification photographs

11. What is the punishment meted to a pilot in command who refuses to submit his blood-alcohol test as requested by a law enforcement officer?

a) Suspension or withdrawal of the remote pilot certificate of the pilot in command

b) The small unmanned aircraft can be withheld for about eight hours or more

c) No severe punishment is attached to such an offence.

12. In the case of an emergency during failure of the standard remote identification, what should be the course of action of the pilot?

a) Safely land the aircraft as soon as possible

b) Send a signal to the authority in charge of control

c) Activate the aircraft's navigation lights.

13. Which Category of small-unmanned aircraft must have an airworthiness certificate issued by the FAA.

a) First category

b) Second category

c) Fourth category

14. If a client has contracted you to conduct a category one operation, which small unmanned aircraft weight is suitable for the operation.

a) The small-unmanned aircraft should be equivalent to or less than 0.55 pounds.

b) The small-unmanned aircraft should be equivalent to or less than 0.65 pounds.

c) The small unmanned aircraft should be equivalent to or less than 0.75 pounds

15. According to the FAA regulation, who is responsible for determining the performance of a small-unmanned aircraft?

a) The remote pilot-in-command of the unmanned aircraft

b) The manufacturer of the small unmanned aircraft

c) The owner of the small unmanned aircraft

16. If the time is currently is 7:00 am and the sunrise at 7:40 am with the assumption that your unmanned aircraft does not have anti-collision lighting. How long must you wait before you can fly?

a) 50 minutes

b) 40 minutes

c) 30 minutes

17. A small-unmanned aircraft causes an accident, and its crew member loses consciousness. When do you report the accident?

a) It is not necessary to report any accident.
b) When requested by the UA owner
c) Within ten days after the accident

18. What are the characteristics of stable air?

a) Good visibility and steady precipitation
b) Poor visibility and steady precipitation
c) Poor visibility and intermittent precipitation.

19. The objective of this pre-flight operation is to determine, from the pilot's point of view ________

a) Check the operating conditions of the aircraft
b) Prevent the occurrence of an accident
c) To fulfil the law of FAA

20. Which of the parts is not included in the external inspection?

a) Right-wing inspection
b) Flap inspection
c) None of the above

21. Wing-end inspection is the inspecting of the _____

a) Inspecting the wings of the UA's
b) Inspecting the movement of the drone

c) None of the above

22. Why is checking the aircraft important during a pre-flight inspection?

a) For proper manoeuvring of the US's
b) To check the hinges slackness
c) None of the above

23. If the propeller is dented, the UA's can be used for commercial purposes

______

a) True
b) False
c) None of the above

24. Which of the following is a common problem in the ventilation tank of the UAS?

a) Dust
b) Lid loss
c) Breakage

25. Which of the following is correct when a pilot is scanning for traffic?

a) Careful observation of the different segments of the air space
b) Concentrate on the movement of the wind
c) Careful observation and detail attention to the information from the ATC

26. To avoid a collision from incoming UA's, a pilot should ensure to _____

a) Pay close attention to nearby UA's
b) Receive and communicate signals to ATC
c) Observe the airspace before piloting the UA

27. According to the FAA, which officer is responsible for processing the remote pilot certificate and the sUAS rating?

a) The officer-in-charge
b) Commercial licenser
c) Remote Pilot in Command.

28. According to the FAA, what is the time interval for recurrent training of a remote pilot?

a) Six-month interval

b) Twelve-month interval

c) Twenty-four-month interval

29. When is it proper to call an aircraft Unmanned Aircraft System (UAS)?

a) When the aircraft is weighing less than 55kg

b) When the aircraft is weighing equal to or less than 55 pounds

c) When the aircraft is weighing above 55 pounds

30. What does an unmanned aircraft mean in simple terms?

a) Aircraft used without any human inputs

b) Aircraft used with human operating it

c) Aircraft without the piloting of humans from within

31. According to FAA, the following are the requirement for obtaining a Part 107 license except for____

a) Commercial purpose

b) Weighing above 0.55 lbs.

c) For fun

32. Which of the following is correct about the usage of the UA's by a licensed pilot?

a) The UA's must be used under the supervision of the Remote officers

b) The UA's can only be used during the daylight period

c) The UA's can be used when permission is granted from the authority in-charge

33. _____ is someone who observes and reports the occurrence of accidents in the different airspace.

a) ATC visual observer

b) The police

c) FAA agents

34. One of the following is an important step before piloting the UAS.

a) Ensure to receive clearance from the ATC
b) Ensure to balance the weight of the aircraft
c) All of the above

35. According to the new regulation of the FAA, a drone can move over a vehicle if ____________

a) Granted clearance from the appropriate authorities
b) If it is under category four of the UA's classifications

c) None of the above

36. A waiver may be granted to a pilot on the following conditions.

a) When the pilot is controlling the aircraft outside the United States of America

b) If the aircraft will not cause any harm to lives and properties

c) None of the above

37. When is the minimum required days needed to present a waiver document?

a) Ninety days

b) Thirty days

c) Ten days

38. Which of the following injuries should a pilot report to the FAA?

a) Small bruises
b) Bone dislocations
c) Λn injury rcquiring an overnight hospital stay.

39. If you crash a property worth five hundred dollars, when should you replace the damaged property?

a) Ten days
b) Ninety days
c) Fifty days

40. Under Part 107, you must cease operating a sUAS at__________

a) Sunset.
b) The end of evening civil twilight
c) The beginning of morning civil twilight

CHAPTER FOUR

Air Space Test Questions

1. What is the purpose of airspace classification?

a) To determine the movements of aircraft, the purpose of operations
b) To cause a division in the airspace
c) For swift movement in the airspace

2. In the United States, the airspace is classified as both_______

a) Special and non-special airspace
b) Controlled and uncontrolled air-space
c) Military and Civilian airspace

3. Which of the following classes is not in the controlled airspace?

a) Class B, C, D, and E airspace

b) Class A and G airspace

c) Class G and Special use airspace

4. Class B airspace has been established to separate ________

a) Incoming and outgoing UA's.

b) All incoming aircraft's

c) All outgoing aircraft's

5. What is the mean sea level (MSL) surface of the class B airspace?

a) 50,000 MSL

b) 30,000 MSL

c) 10,000 MSL

6. Communication with pilots is mandatory in Class B airspace.

a) True

b) False

c) None of the above

7. Class C airspace areas are designated to specific ______

a) City

b) Armed spots

c) airports where ATC exists

8. The class C airspace surface is from ______ elevation.

a) 20,000 feet

b) 4,000 feet

c) 10,000 feet

9. The inner-circle has a radius of _____ and begins at ground level.

a) 5 Nautical Miles

b) 10 Nautical Miles

c) 15 Nautical Miles

10. Which of the class of airspace is not under the control of the ATC.

a) Class A

b) Class B

c) Class G

11. A small-unmanned aircraft planing to operate within Class G airspace should_______.

a) Take permission from ATC

b) Be vigilant not to cross over

c) Minimum weather conditions are required

12. What type of airport is Pueblo Airport?

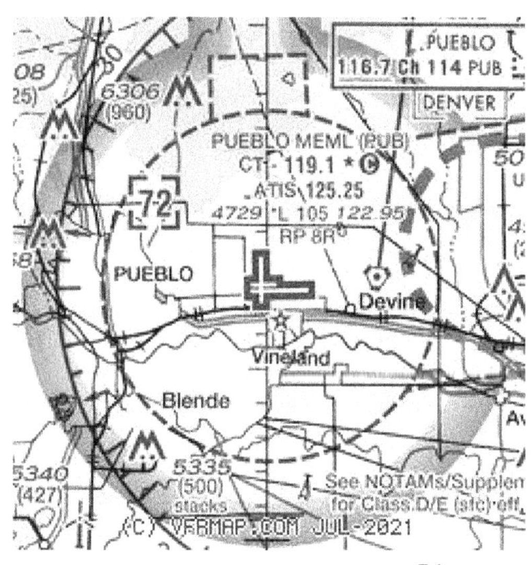

a) Class B

b) Class C

c) Class D

13. What is the height of the Class E AIRSPACE?

a) 14,500 feet MSL

b) 1937 feet MSL

c) None of the above

14. The remote pilot in command of a UA that is planning to operate within Class C airspace must _______

a) Ensure to use a visual observer

b) Ensure to submit the plan of the flight

c) Ensure to receive clearance from ATC

15. The NALF (Naval Auxiliary Landing Field Fentress) airport is in _______ airspace.

a) Class D
b) Class G
c) Class C

16. Supposing a farmer had employed your services to survey his crop farm located in Devil's Lake West MOA, east of the area. What strategy will you employ to locate the activeness of the MOA?

a) Consult care centre for their help

b) Access the information via the US database

c) Access the information in the directory of military operations

17. Departures from airports without a control tower must establish communication with ________

a) The US Tower

b) The Air force tower

c) The ATC

18. Which of the airspace has numerous amounts of airspace?

a) The class D

b) The class A

c) The class E

19. Which airspace is present both in the controlled and uncontrolled region?

a) The special use airspace
b) The military airspace
c) The prohibited airspace

20. The prohibited areas are established for what purpose?

a) Recreational Purpose
b) Security Purpose
c) Commercial purpose

21. The restricted areas include airspace where flying operations are prohibited.

a) True

b) False

c) None of the above

22. Which of the airspace usually carries missiles and aerial artillery?

a) Prohibited areas

b) Dangerous areas

c) Restricted areas

23. The purpose of the dangerous area is to ______________

a) Warn non-participating pilots of the potential danger

b) Safeguard the ATC

c) Prevent international hazard

24. Which of the areas on the map is shown in letters?

a) Class A
b) Dangerous area
c) Alert areas

25. _______ is the floor space of the class C savannah airspace.

a) It is below the 1300 ft AGL
b) It is at 1300 ft MSL
c) It is 1700ft MSL

26. Which instruction is necessary for the pilot in charge who wants to operate the UAS within the class C?

a) The pilot must utilize the visual observer instrument
b) The pilot in charge should come up with a plan
c) The pilot in charge must receive an authorization

27. During the pre-flight period, where can you find information about planning to mount an R-2305?

a) You can find the information in the aeronautical manual
b) You can find the information in the US charts of supplements
c) You can find the information under special request

28. In the pueblo airport, a magenta line is present in the circle; what is it for?

a) It represents the class E

b) The line can be described as an isogenic line

c) The line represents that this region is a secure area

29. In pueblo airport, ________ is present on the map?

a) a magenta line

b) a wide circle

c) All of the above

30. The requirement in terms of visibility for an unmanned aircraft

when it is close to the plantation airport is _______

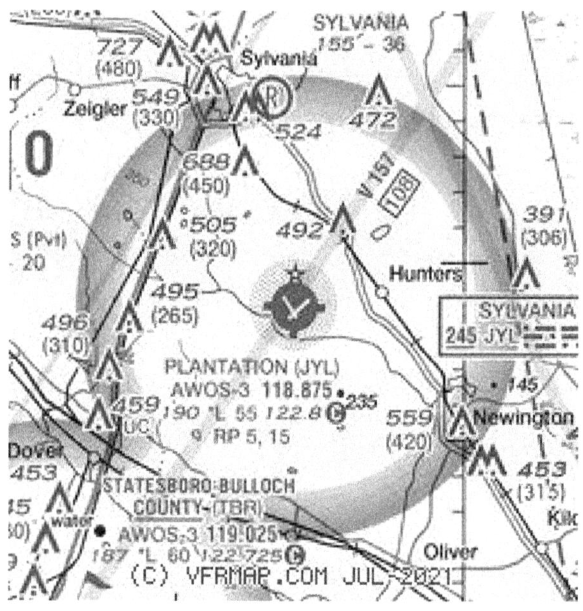

a) Ten statute miles

b) Fifteen statute miles

c) Three statutes miles

31. On the map of Lake Drummond, there is a small flag; what is the aim of the flag?

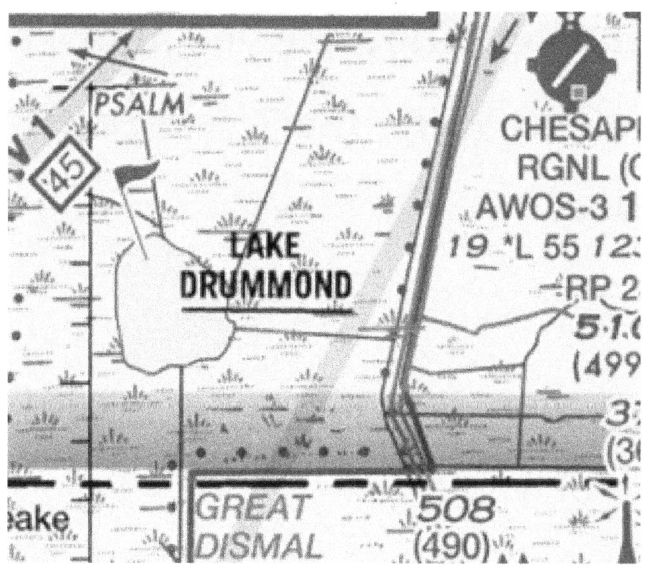

a) The aim is to provide a check for the VFR and the traffic present in the airspace

b) It serves as a GPS that can direct both the manned and unmanned aircraft

c) It is a significant obstruction for redirecting aircrafts

32. In what condition are you allowed to fly the UAs when there is a restriction order?

a) When there is an FAA license

b) When there is evidence for a certificate of waiver

c) When there is evidence of MOA

33. What is the floor of the Class B airspace at Dallas Executive?

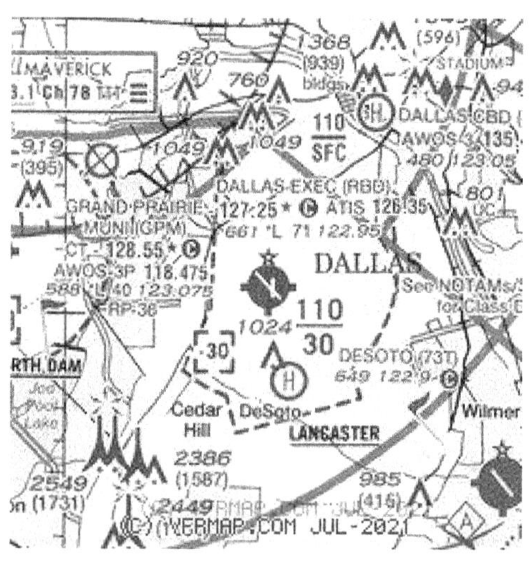

a) At 5000 AGL

b) At 4000 ft MSL

c) At 3000 ft MSL

34. Before entering Class D airspace, radio communication must be established and maintained with the control tower?

a) True

b) False

c) None of the above

35. The Class E does not include airspace within ___________

a) 1400 ft MSL

b) 1450 ft MSL

c) 1,500 ft MSL

36. In Class G airspace, both the pilot and the aircraft must be qualified for an IFR flight.

a) True

b) False

c) None of the above

37. In a class B airspace, _____ is required when flying within a 30NM radius of the primary class B airports.

a) Transmitter

b) Visual observer

c) Transponder

38. What is the major requirement needed to fly in a class A airspace?

a) Endorsement letter from your instructor
b) At least a commercial pilot certificate
c) Part 107 license + UA's ratings

39. The first action needed before flying through a class B VFR corridor is______.

a) Request for a flight plan
b) Call the ATC tower for confirmation
c) None of the above

40. Which of the following is the reason for issuing a temporary flight restriction?

a) The region where major sporting activities are carried out
b) The region where the president and other governmental authorities are travelling
c) A and B

CHAPTER FIVE

Part 107 Weather Test

1. During the weather report, it was forecasted that the ceiling would be 700ft. As an unmanned pilot, what should be the course of action?

a) You will fly at 700 ft AGL

b) You will fly at 800 ft AGL

c) You will fly at 200 ft AGL

2. Low density has several effects; what is its effect on the propeller?

a. The low density increases the propeller efficiency

b. The low density decreases the propeller efficiency

c. Propeller efficiency is standard despite the low density

3. According to the FAA, _________ is stipulated visibility of unmanned aircraft.

a) Five statute miles

b) Ten statutes miles

c) Three statutes mile

4. Different international airports have different wind speed and wind direction; Memphis International Airport has _______ wind source?

a) The wind source is from the south-southwest, 12 knots

b) The wind source is from the north-west, 30 knots

c) The wind source is from the south-west, 3 knots

5. The weather condition when flying through the Los Angeles International Airport has been observed to have ______.

a) Presence of fog and strong wind from the SW

b) Absence of moisture and mild wind from NE

c) Sparse clouds at 700 feet, the temperature of 16 °C

6. During pre-flight, you observe that the weather is moist and unstable; what characteristics are you based on?

a) Absence of clear air and cool wind
b) Presence of turbulent and drizzling rain
c) Presence of smoke in the weather

7. Aeronautical meteorological data must be distributed to users immediately.

a) True
b) False
c) None of the above

8. Surface Aviation Weather Observations consist of the following except?

a) Government officials

b) Meteorologist

c) Weather reporters

9. Surface Aviation Weather Observers provide the following information about weather conditions.

a) Temperature/dew point

b) Reports of UAs

c) None of the above

10. Which of the following is not a meteorological instrument used to measure the weather condition?

a) Barometer

b) Measuring cylinder

c) Wind vane

11. The measurement of the pressure is essential in a drone.

a) To set the altimeter of the drone

b) To properly calibrate the weight of the drone

c) None of the above

12. The mercury barometer consists of a tube about _______

a) 950mmHg

b) 850 mmHg

c) 500mHg

13. The mercury thermometer registers the ________

a) The relative humidity

b) The wind speed

c) The temperature

14. The psychrometer consists of ________ bulb thermometers

a) A pair of glass mercury

b) Water in the glass

c) None of the above

15. Which instrument is used to measure the sky conditions?

a) Thermometer

b) Psychrometer

c) Ceilometer

16. _______ is used to determine the intensity of the precipitation

a) Pluviographs

b) Ceilometer

c) Psychrometer

17. If you decided to observe the automatic precipitation sensor,which liquid does it measure?

a) Water

b) Ice

c) All of the above

18. For the ATC, the meteorological instrument is the ___________

a) Remote sensor

b) Remote control

c) Observation glass

19. The scatterometers are instruments that measure ________

a) Intensity of light

b) Amount of light

c) Scattering of light

20. The Transmissometers are used to measure the ______

a) Runway visual range

b) Current

c) None of the above

21. Suppose there is a heavy thunderstorm in an environment; what action should an unmanned pilot carry out?

a) Maintain the precipitation range from the drone

b) Quick landing

c) None of the above

22. As a pilot of an unmanned aircraft, what is the best use of a Weather Prognostic Charts?

a) For visual planning before embarking on a journey
b) For proper analysis of the weather condition
c) For determining the area to avoid when flying

23. The air can hold a certain amount of water; what factor does it depend on?

a) It depends on the dew point
b) It depends on the air temperature
c) It depends on the air

24. Dew point simply mean________?

a) The temperature at which condensation and evaporation are equal.

b) The temperature at which dew will always form.

c) The temperature to which air must be cooled to become saturated.

25. In which environment is aircraft structural ice most likely to have the highest accumulation rate?

a) Cumulus clouds with below-freezing temperatures.

b) Freezing drizzle.

c) Freezing rain.

26. For aviation purposes, the ceiling is defined as the height above the Earth's surface of the

a) Lowest reported obscuration, and the highest layer of clouds reported as overcast.
b) Lowest broken or overcast layer or vertical visibility into an obscuration
c) The lowest layer of clouds was reported as scattered, broken, or thin.

27. _______ is the stage at which thunderstorms will attain their greatest intensity.

a) At maturity stage
b) At cumulus
c) At the downdraft stage.

28. When is it advised to fly small-unmanned aircraft?

a) When the weather is cleared
b) When clearance has been given from FAA and ATC
c) When the necessary instrument for piloting the small unmanned air-craft is available

29. Which of the following is correct?

a) Consulting the weather report is not necessary before piloting the aircraft
b) The weather forecast may be wrong occasionally

c) Flying the classification C does not involve much interference from weather.

30. Weather changes often appear in ________ quickly.

a) Class A
b) Class C
c) Class G

CHAPTER SIX

Loading and Performance

1. If your aircraft's centre of gravity is tilted backward, what is the likely result?

a) The aircraft will have difficulty recovering from a stall position.

b) The aircraft will not be able to maintain a constant turn.

c) The aircraft will have higher air-speed.

2. When piloting the drone, you observed the temperature of the air is outside air is hot compare to the

temperature of the drone, then the altitude is _______

a) The high temperature from the altitude
b) Low temperature from the altitude
c) The average temperature from the altitude

3. Suppose you are to pilot an unmanned aircraft that weighs about 20Ibs, to make a 60^0 turn, what should be the aircraft's structural weight.

a) 40Ibs
b) 60Ibs
c) 90Ibs

4. Which among the following factor is more likely to increase the density altitude of an airport?

a) When there is an increase in the pressure

b) When there is an increase in the temperature

c) When there is an increase in the relative humidity

5. How can you find the centre of gravity of UAS?

a) Multiplication of the weight of the UAS

b) The weight of the UA's is divided by the total moment

c) The total moment is dividing the weight of the UAS

6. When will altitude and density become equals?

a) At 0^0 F (sea level)

b) At standard temperature

c) When there is no error in the altimeter

7. Who determines the performance of the UAS?

a) The remote piloting officer

b) The manufacturer of the UAS

c) The commercial man

8. The main cause of the GC's lateral displacement is ________

a) Imbalance of wings
b) Overweight
c) None of the above

9. To alleviate GC's lateral displacements, it is normally recommended to balance the _______

a) Remove heavyweights from the drone
b) Balance the weights on both sides regularly
c) None of the above

10. The Center of Gravity (CG) is the point where the force of ____________

a) Gravity moves the object

b) Gravity stabilizes the object

c) Gravity affects the body mass

11. Is CG a fixed point?

a) Yes

b) No

c) None of the above

12. Which of the following is not a negative consequence of overweight?

a) The maximum achievable distance is longer

b) Braking capacity is reduced

c) The cruising speed is lower

13. The first warning of overweight is ______

a) Failure in the control system
b) Poor performance
c) Sliding wings

14. At what point are the forces acting on the UAS equilibrium?

a) During acceleration of the UA
b) During unaccelerated flight
c) When the aircraft is at rest on the ground.

15. The path followed by an aircraft during its movement in the sinus of the air is called the flight path.

a) True

b) False

c) None of the above

16. The leading edge is ______

a) The rear part of the wing

b) The front or front part of a wing profile.

c) The front part of the engine

17. The upper curvature that goes from the leading edge to the edge of output on a wing is called?

a) Rope.

b) Extrados.

c) Intrados.

18. It is called the centre of pressures, to the point where:

a) The wings have more pressure.
b) The pilot must be in place.
c) The resultant of the aerodynamic forces are applied.

19. The equidistant line between the upper surface and the lower surface of a wing profile is called_________?

a) Average curvature.
b) Thickness.
c) Wingspan.

20. The lift is ____________

a) The upward force is perpendicular to the relative wind and developed to support the weight of the plane.
b) The aerodynamic force, plus the parasitic drag
c) The force is perpendicular to the relative wind and developed to perform the traction of the plane

21. When is the weight of the aircraft balanced?

a) When the force is equal to zero
b) When the upward force exceeds the downward force
c) None of the above

22. Which of the following is correct about loading the aircraft?

a) The aircraft load affects the performance

b) The load can affect the condition of the pilot

c) Overweight is advised to ensure the aircraft land

23. Which of the following will affect performance more?

a) Weight

b) The altitude of the aircraft

c) Pilot experience

24. At what point are the forces acting on the UA's at equilibrium?

a) During acceleration of the UA
b) During unaccelerated flight
c) When the aircraft is at rest on the ground

25. Recreational drones are not allowed to fly during bad weather.

a) True
b) False
c) None of the above

26. The main cause of the GC's lateral displacement is ____

a) Imbalance of wings
b) Overweight
c) None of the above

27. Weather conditioning is caused by __________?

a) Solar radiation

b) Precipitation

c) All of the above

28. Which of the following is the major causes of the drone accident?

a) Overweight

b) Tiredness

c) Experience

29. Low density has several effects; what is its effect on the propeller?

a. The low density increases the propeller efficiency

b. The low density decreases the propeller efficiency

c. Propeller efficiency is standard despite the low density

30. _______ is the reason for not flying when the CG is above the normal.

a. It affects maneuverability

b. It affects the wings of the aircraft

c. All of the above

CHAPTER SEVEN

Part 107 Operations Test

1. The meaning of the sign below confirm ________

a) The runway is 22

b) The routing runway is 22

c) The taxiway is 22

2. Which among the option is correct in identifying a military air station?

a) Presence of a standing flag (red and yellow color)

b) Presence of small flash (white and green)

c) Presence of a tall flag (green, yell-ow) and a flash

3. To indicate the wind cone segm-ented circle, which of the runway is needed _____

a) Right-hand traffic on Runway 9

b) Right-hand traffic on Runway 18

c) Left-hand traffic on Runway 36

4. When is it necessary for unmanned aircraft to carry out maintenance procedures?

a) In the absence of maintenance procedures from the manufacturers

b) In the occurrence of an accident

c) There is no need for any maintenance procedure

5. Suppose in the absence of ATC broadcast, which other methods can be used to indicate the weather condition?

a) The weather condition and its visibility

b) VFR sky weather readings

c) None of the above

6. What does the letter A represent in an airport?

a) Landing of unmanned aircraft

b) Take-off point of unmanned aircraft

c) None of the above

7. According to ______ is responsible for the maintenance of an unmanned aircraft performed.

a) Pilot in-charge

b) Manufacturers

c) None of the above

8. If an unmanned aircraft is powered by lithium batteries, what is the prescribed precaution to prevent an accident?

a) Keep the battery from water or any other liquid

b) Ensure to charge the battery before usage

c) All of the above

9. When flying the unmanned aircraft, suppose a bird is hurt during the remote piloting; what should be the remote-pilot-in charge action?

a) Report to the FAA
b) Report to the local community
c) None of the above

10. The ATIS carries the function of broadcasting continuous recorded information of the _______

a) The pilot whose radar is in dangerous proximity to terrain or an obstruction

b) The pilot that needs non-essential information

c) None of the above

11. While piloting the unmanned aircraft, which of the classes necessitates two-way radio communication?

a) Class A

b) Class C

c) Class E

12. The Sign below means _____.

a) It confirms to the pilot that they are in the taxiway B location

b) It is a warning sign to the pilot about the taxiway

c) All of the above

13. Is a two-way communication necessary with the ATC before landing or take-off when piloting the unmanned aircraft?

a) True

b) False

c) None of the above

14. Which among the following is the consequences of a damaged lithium battery?

a) The result to fire accident

b) The result of the change in the CG of the UA

c) Reduces the speed of the aircraft

15. What should be the cause of action from the operator of a small-unmanned aircraft, if the manufactured failed in its duties of maintenance before purchase?

a) Report to the FAA

b) File for damages in the law court

c) The operator should schedule a maintenance operation for the aircraft

16. Tiredness or stress is one of the most often cited factors involved in aviation accidents.

a) True
b) False
c) None of the above

17. Fatigue is detrimental for the unmanned pilot because ______

a) It prevents the pilot from giving maximum performance
b) It inhibits pilot actions in cases of emergency
c) All of the above

18. The slightest alcohol consumption can impair judgment.

a) True
b) False
c) None of the above

19. Peradventure a pilot consume alcohol, he should not fly a drone till _____

a) After 24-hours
b) After 18-hours
c) After 8-hours

20. How often does the drone pilot need to inspect their drone to ensure it is in good working condition?

a) Before each flight
b) Daily
c) Monthly

21. Which of the following is correct about the operation of a small-unmanned aircraft?

a) It requires skill
b) It requires experience
c) None of the above

22. What is the maximum allowable altitude when flying a drone?

a) 400 ft
b) 500 ft
c) 800 ft

23. What is the maximum speed required for the above altitude?

a) 100 mph

b) 250 mph

c) 80 mph

24. According to FAA, what is the maximum speed for a civilian drone?

a) 100 mph

b) 80 knots

c) All of the above

25. Is it proper to fly in restricted airspace during emergency?

a) Yes

b) No

c) None of the above

26. Is it allowed to fly above people?

a) Yes

b) Except with a permit

c) No

27. The acting force on the drone is ______

a) Gravitational force

b) Thrust

c) All of the above

28. Can FAA track all the drones in the airspace?

a) Only commercial drones

b) Both commercial and recreational drones

c) None of the above

29. The ATIS carries the function of broadcasting continuous recorded information of the ______

a) The pilot whose radar is in dangerous proximity to terrain or an obstruction
b) The pilot that needs non-essential information
c) None of the above

30. ESC means ______

a) Escape
b) Electronic Speed Controllers
c) End Soaring Control

CHAPTER EIGHT

Part 107 Risk Management and Emergency Procedures Test

1. To prevent accidents in the unmanned aircraft system, which is necessary?

a) A risk management course

b) Flying classes

c) All of the above

2. When fitting CRM concepts to the operation of a small UA, _____ must be used in combination.

a) the flight portion only

b) all phases of the operation

c) the communications only

3. A remote pilot who tends to impress others may likely suffer ______

a) Destruction of the engines
b) High collision with other objects
c) Macho

4. Choose the best answer from the following?

a) Little quantity of alcohol can influence the decision making of the pilot
b) Little quantity of alcohol increases the insight of the unmanned pilot decision

c) Consumption of a large quantity of water can destroy alcohol present in the body

5. Which of the following is the most common hazardous attitude exhibit-ted by the remote pilot?

a) Lack of risk management technique and Stress management

b) Resignation and Anti-authority

c) Snap judgement and anti-social behav-iors

6. An unmanned pilot must recognize and understand ____ and _____

a) Signals and Lighting

b) Signs and Inscriptions

c) Maps and weather reports

7. An unmanned pilot must identify and avoid specific terrain.

a) True

b) False

c) None of the above

8. An unmanned pilot must identify and avoid terrible weather conditions.

a) True

b) False

c) None of the above

9. An unmanned pilot must avoid ______

a) Maneuvering

b) Collisions

c) Straight movement

10. What is an inflight emergency?

a) Decisions are taken during emergency

b) Flight control during the accident

c) All of the above

11. One of the following does not involve the risk of piloting small-unmanned aircraft.

a) Collision

b) Invasion

c) Aerial surveillance

12. What should be a pilot reaction if he collides with a strange object?

a) Change the aircraft flight mode

b) Turn of the aircraft engine

c) Activate the return home button

13. Which of the following action is correct during drone flyaway?

a) Put on the GPS

b) Communicate your location

c) All of the above

14. What do you do if you lose your drone?

a) Report to the FAA

b) File a case to the court

c) Summon a search team

15. Which of the following is correct about the procedure for searching for a lost drone?

a) Gather information about the last location

b) Check the predetermined destination of the drone
c) None of the above

16. Is drone manual necessary before piloting small unmanned aircraft?

a) Yes
b) No
c) None of the above

17. What happens when the drone in airspace goes out of range?

a) Loss of signal
b) Loss of control
c) All of the above

18. Is it permitted to fly a drone at night?

a) Yes

b) No

c) Only with permit

19. _____ is needed during a flight review.

a) Review of the current general operating and flight rules

b) A review of those maneuvers and procedures

c) All of the above

20. According to FAA, is it permitted for the drone to fly when it is rainy?

a) Yes

b) No

c) None of the above

Answers to Part 107 Regulations Question

1. a
2. a
3. c
4. a
5. a
6. b
7. a
8. b
9. a
10. b
11. a

12. a

13. c

14. a

15. a

16. b

17. c

18. a

19. a

20. c

21. a

22. b

23. b

24. a

25. c

26. b

27. c

28. b

29. b

30. c

31. c

32. c

33. a

34. c

35. a

36. b

37. c

38. c

39. a

40. b

Answer to Air Space Test Questions

1. a

2. b

3. c

4. a

5. c

6. a

7. c

8. b

9. a

10. c

11. c

12. c

13. a

14. c

15. c

16. b

17. c

18. c

19. a

20. b

21. a

22. c

23. a

24. c

25. b

26. c

27. b

28. c

29. a

30. c

31. a

32. b

33. c

34. a

35. c

36. a

37. c

38. c

39. c

40. c

Answers to Part 107 Weather Test

1. c

2. a

3. c

4. a

5. c

6. b

7. a

8. c

9. a

10. b

11. a

12. b

13. c

14. a

15. c

16. a

17. c

18. a

19. c

20. a

21. b

22. b

23. a

24. c

25. a

26. b

27. b

28. b

29. b

30. b

Answers to Loading and Performance Test Question

1. a

2. a

3. a

4. c

5. b

6. b

7. a

8. a

9. b

10. c

11. b

12. a

13. b

14. c

15. a

16. b

17. c

18. c

19. a

20. a

21. c

22. a

23. c

24. c

25. b

26. a

27. c

28. a

29. b

30. c

Answer to Part 107 Operations Test

1. a

2. c

3. b

4. a

5. a

6. b

7. b

8. c

9. a

10. a

11. b
12. c
13. a
14. a
15. a
16. a
17. c
18. a
19. c
20. a
21. a
22. a

23. a

24. c

25. b

26. b

27. b

28. a

29. a

30. b

Answers to the Part 107 Risk Management and Emergency Procedures Test Questions

1. c

2. c

3. b

4. a

5. b

6. a

7. a

8. a

9. b

10. a

11. c

12. a

13. a

14. a

15. c

16. a

17. c

18. a

19. c

20. b

About the Author

Robert Gonzalo is an FAA Part 107 certified remote pilot and a licensed aviator. He has spent a lifetime piloting and operating unmanned aircrafts, and has conducted a variety of training to a number of astute drone pilots in the industry.

Robert holds a Bachelor of Science Degree in Aviation Management, from the Ohio State University, Columbus, Ohio.

www.ingramcontent.com/pod-product-compliance
Ingram Content Group UK Ltd.
Pitfield, Milton Keynes, MK11 3LW, UK
UKHW022004190726
13853UKWH00004B/1716